HOW TO MANAGE YOUR KINDLE

The Ultimate Guide for Complete Novice On How to Manage Your Kindle Including Manage Books and Docs, Manage Subscriptions, Update Kindle Payment Detail

I0492856

BY

JAMES R. HALE

Copyright©2018

COPYRIGHT

Charles S. Mills

TABLE OF CONTENT

CHAPTER 1

INTRODUCTION

Amazon's Manage Your Kindle guide is a final bus stop point for managing your Kindle content also with your Kindle device. The manage Your Kindle page is more helpful, if you have several Kindle devices.

Manage Your Kindle page can be used to send books from your Kindle library to any of your Kindle devices. Also, it can be use to see the periodicals you subscribe to, and the subscriptions can be managing as well. Links help you to Manage your method of payment to Amazon so that items you buy on your Kindle Fire will get charged to the right credit card. Lastly, you can register and deregister Kindles and rename your devices from the Manage Your Kindle page.

This guide gives the breakdown how you can Manage Your Kindle with step by step process even as a novice, all you have to do is just follow the steps as instructed. But if you find this very difficult, there is no needs to worry because I got you cover.

All over the world millions of kindle users haven't been able to know how they can Manage their kindle, but this book gives the breakdown of all solution to any problem you might encounter. With proper medication and better understanding you will have the knowledge of how to Manage Your Kindle.

Thankfully each steps are very easy and simple to follow, that even a beginner can master it in a few minutes.

CHAPTER 2

HOW TO MANAGE YOUR KINDLE

In this guide, you will be show how you can use Amazon's Manage Your Kindle page to keep track of your subscriptions and books and handle your payment and device information.

Manage content

Manage your library

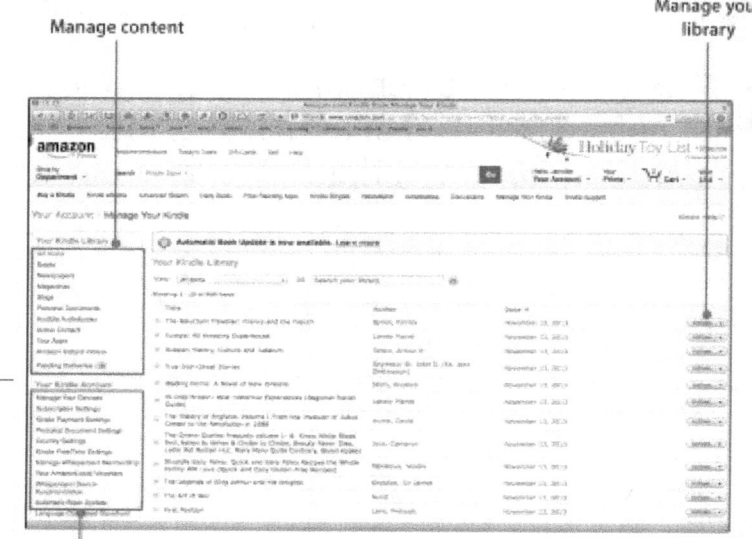

In order to Manage Your Kindle, the following topics include:

a. Managing your books and docs.

b. Managing subscriptions.

c. Updating Kindle payment information.

d. Managing your Kindle devices.

Amazon's Manage Your Kindle page is a final bus stop location for managing your Kindle content also with your Kindle device. The manage Your Kindle page is more helpful, if you have more than one Kindle device.

Manage Your Kindle page can be use to send books from your Kindle library to any of your Kindle devices. Also, it can be use to see the periodicals you subscribe to, and the subscriptions can be manage as well.

Links help you to Manage your method of payment to Amazon so that items you buy on your Kindle Fire will get charged to the right credit card. Lastly, you can register and deregister Kindles and rename your devices from the Manage Your Kindle page.

MANAGING BOOKS AND DOCUMENTS

All books and documents can be view in your library using Manage Your Kindle. They can be transfer to your Kindle. Ebooks you bought from Amazon's Kindle Store appear in the books content library, while ebooks you bought from a source other than Amazon will be seen in the Personal Document content library. Docs that will appear in Manage Your Kindle have been emailed to your Kindle.com email address for document conversion. Manage Your Kindle doesn't list docs that you load to your Kindle Fire using the micro-USB cable (this process is known as side loading).

What's Up with Docs?

When I talk about "docs," I'm referring to Kindle Personal Documents. I sue the term **docs** because it is how the Kindle Fire refers to that content library on the device. However, The Manage Your Kindle page on the Amazon site, uses the term Personal Documents.

Chapter 3, "Managing Your Personal Documents and Data," gives all about **docs** in detail.

ACCESSING MANAGE YOUR KINDLE

Manage Your Kindle is a web page which you access by making use of the web browser on your computer or you can use the silk on your Kindle Fire.

a. Right in your web browser, head to www.amazon.com.

b. Hover over the "Your Account Drop-Down Menu" if the Amazon site is being access from the Silk browser on your Kindle Fire, then hit instead of hovering.

c. Select Manage Your Kindle. If prompted, log in and input your Amazon password and email address.

A Quicker Way to Manage Your Kindle

You can arrive at the Manage Your Kindle page directly by heading to www.amazon.com/manageyourkindle in your web browser.

VIEWING BOOKS AND DOCS

The Manage Your Kindle default view lists all the books, apps, subscription, magazines and private docs on your Kindle devices. The books and docs are cover in this section, and I talk about managing magazines and newspaper in the next section.

However, from the Manage Your Kindle page, you can choose the kind of content you want to view by using the view drop-down list. You can find content by title, author, or date by hitting one of the column headers. The first hit of a column header comes in descending order, hitting the same column header again put that column in ascending order.

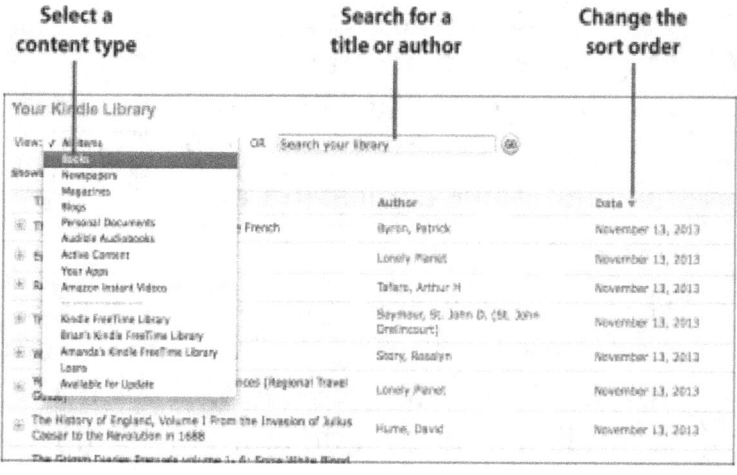

Select a
content type

Search for a
title or author

Change the
sort order

If you plan to view details on an item, hit on the plus sign beside the item title. If you own a lot of content and you want to look for a particular item, input a search term and press GO.

Why Use Manage Your Kindle?

Many functionalities in Manage Your Kindle, such as transferring a book from the Amazon cloud to your device, can be accomplished directly on your Kindle Fire. However, if you decide to borrow a book to another Kindle user or permanently terminate a book from your Kindle library, you complete those tasks through Manage Your Kindle. It's a suitable method to deliver content to several Kindle devices or apps on your account or when someone else in your house is using your Kindle Fire you can manage your account.

SENDING BOOKS AND DOCS TO YOUR KINDLE

You are able to send books and docs to a Kindle app for Android, iPhone, iPad, and iPod Touch. Books can be send, but not docs, to the other Kindle apps. Content is delivered within a minute, supposing you are connected to Wi-Fi.

Kindle Apps

Mentioning Kindle apps in this chapter, I'm not referring to apps installed on your Kindle Fire. But rather I am referring to the Kindle app that you can use on a smartphone device, computer, or tablet to read Kindle ebooks.

a. Search for the book or doc that you wish to send to your Kindle.

b. Hover over (or click on the silk browser) the action drop-down list

c. Hit Deliver to My.

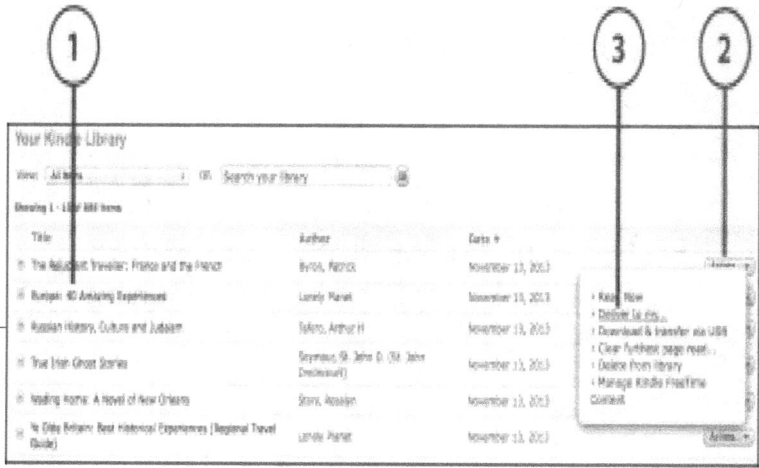

d. Choose the device from the drop-down list. But if the doc you want to send to your Kindle is not in any format supported by a particular device, it indicates that the device isn't available in the drop-down list.

e. Hit Deliver.

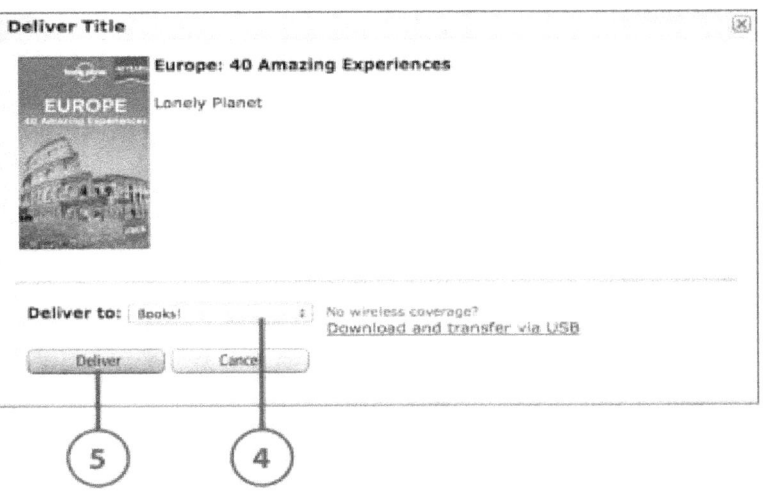

GET BOOKS DOWNLOAD TO A COMPUTER

Books can be download to your computer, but not docs. After you download a book, using a micro-USB cable you can side load it to your Kindle Fire.

a. Hover over (or click) the Action drop-down list.

b. Tap download and then transfer via USB.

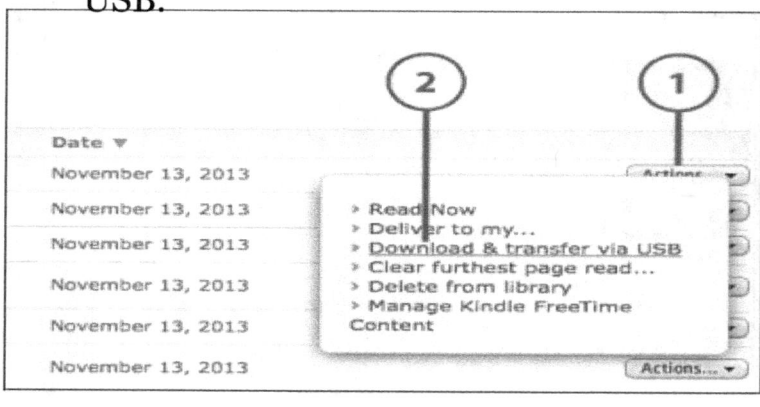

c. Choose the Kindle you want to transfer the book to.

d. Tap download and use the browser download option to save the book.

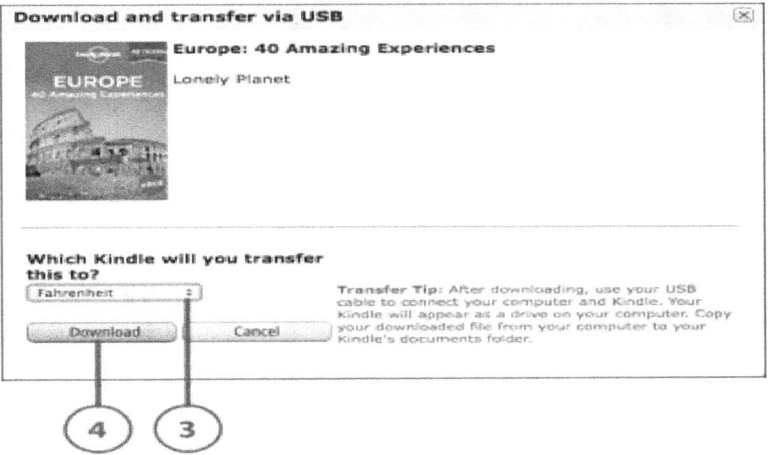

DELETE BOOKS AND DOCS

Books and docs can be deleted from your library if you don't want to read them again. Ensure to use this feature with caution because doing so takes away the item forever. If you delete a book you bought from Amazon, you have to buy it again if you change your mind.

a. Point to the Actions drop-down list.

b. Tap "Delete from Library".

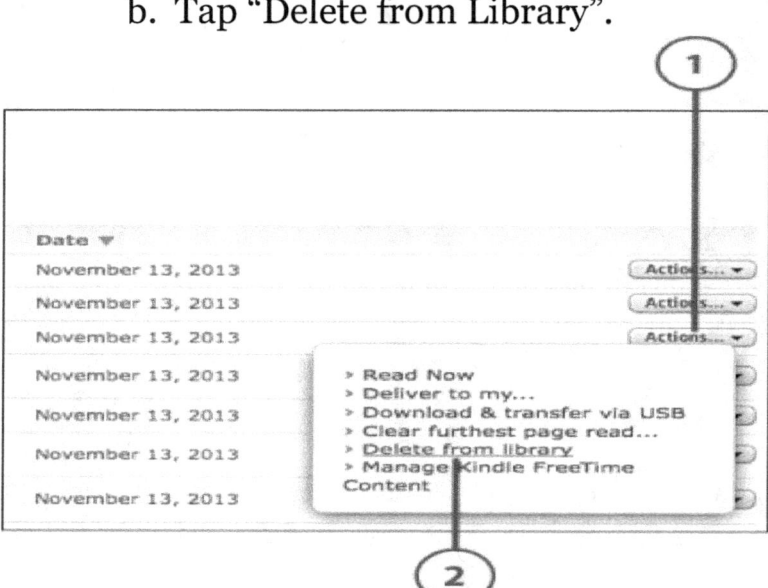

c. Tap "YES" to confirm that you want to delete the book from your library forever.

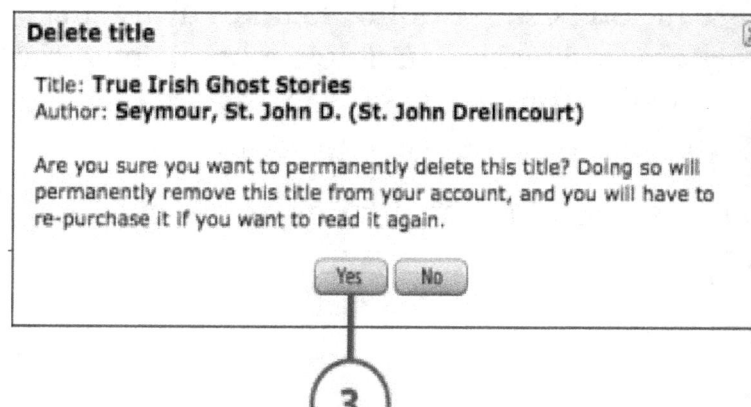

Charles S. Mills

CHANGE YOUR KINDLE EMAIL ADDRESS

You can use your Kindle email address to send docs straight to your Kindle. On Manage Your Kindle page, you can change the email address for your Kindle Fire.

a. Tap "Personal Document Settings".

b. Tap "Edit" beside the Kindle email address you plan to change.

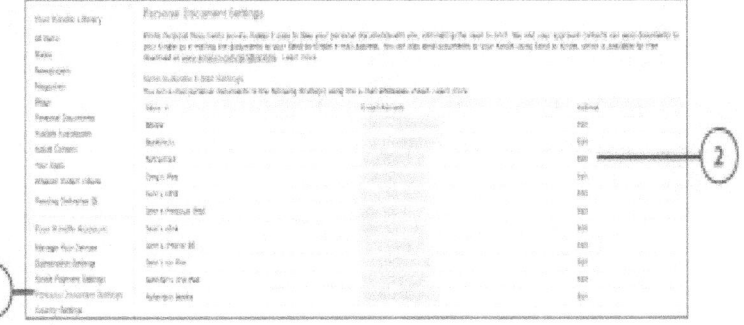

c. Input the new email address.

d. Tap "Update".

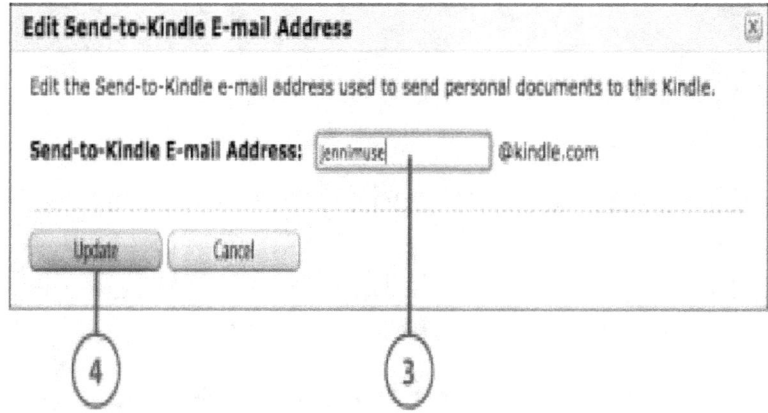

INCLUDING A RATIFIED EMAIL FOR DOCS

To stop spam on your Kindle device, Amazon delivers only docs emailed from a ratified list of senders. Using Personal Document Settings, you can add a ratified email address.

a. From Personal Document Settings, tap "Add a New Approved Email

Address" in the Approved Personal Document Email List section.

b. Input the email address you plan to approve. You can also input a partial email address, such as @yourcompany.com, to permit all senders from that specific domain.

c. Tap Add Address.

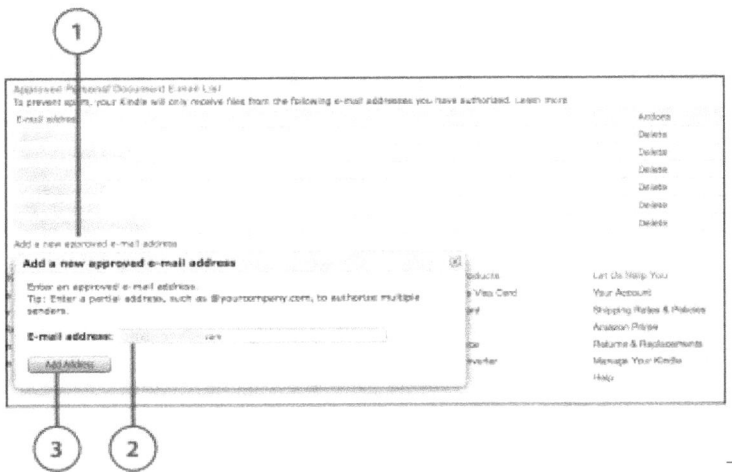

Deleting a ratified email address

By tapping "Delete" to the right of the email address on the Approved Personal Document Email List, you can delete an approved email address.

Further More: Putting Your Kindle Email Address to Work

Most non-Amazon online bookstores, like www.omnilit.com, deliver bought items directly to your Kindle account if you give them your Kindle email address. You should ensure to add these providers to your approved email list and keep in touch with the bookstore site's instructions about including your Kindle address to your bookstore account before doing a purchase.

DISABLING DOC ARCHIVING

Docs that are sent to your Kindle are also save in your Kindle library, by default.

Amazon provide you with 5GB of space for personal doc archiving. However, archiving of personal docs can be disable.

a. From Personal Document Settings, tap on "Edit" in the Personal Document Archiving section.

b. Unchoose the box to disable archiving.

c. Tap update.

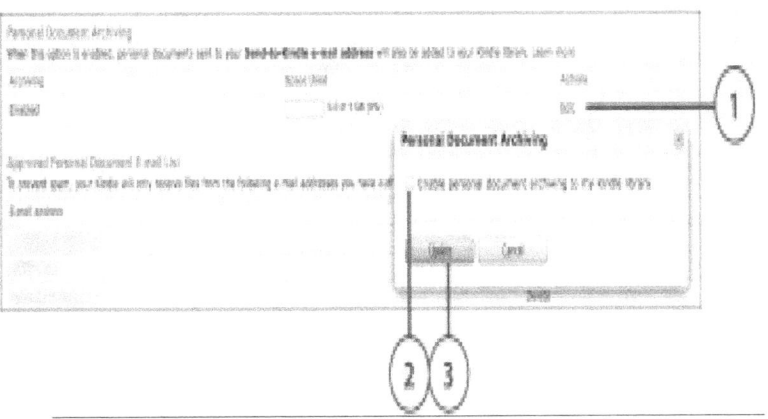

Double The Docs Space

The 5GB personal doc storage in your Kindle library is different from the 5GB of storage on your Amazon cloud drive. Which can also be used for docs (amidst other files). It is advising that you save the kindle personal doc space for ebook bought from other bookstores because it is accessible by email. To save your other personal documents, use your Amazon Cloud Drive. By this it gives the benefit of keeping all your real personal docs together on the Amazon Cloud Drive, making them so easy to organize and manage.

CHAPTER 3

MANAGING YOUR SUBSCRIPTIONS

From the Manage Your Kindle page your magazines and newspaper subscriptions can be manage. You can select which device automatically get your subscription, send previous problems to your Kindle Fire, or download previous problems to enable you sieload them to your Kindle Fire. Lastly, you are able to cancel your subscription completely. ***To know more about subscribing to periodicals, head to chapter 4.***

CHANGING WHERE A SUBSCRIPTION IS DELIVERED

When you first subscribe you can select which device accepts the automatic delivery of subscription content. It automatically accepts the subscription if you subscribe from your Kindle Fire. You can change that choice from Manage Your Kindle. If you have several registered Kindle device this option is available.

a. Tap Subscription Settings.

b. Tap "Edit" for the subscription you plan to change.

c. Choose a device to which fresh editions should be delivered.

d. Tap update.

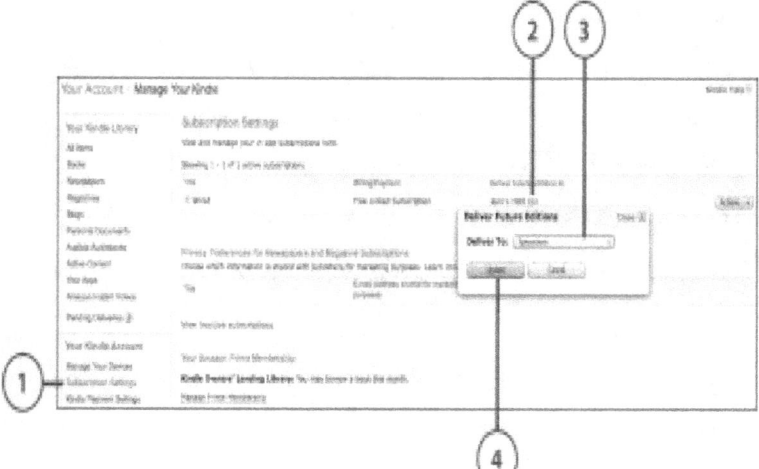

GETTING A SUBSCRIPTION CANCEL

If the magazine reaches your Kindle Fire on a distinct distribution timetable, subscriptions are automatically charged on a monthly basis. If you plan to cancel a subscription, it can be done from Manage Your Kindle.

a. From subscription settings, tap "Actions" for the subscription you plan to cancel.

b. Tap Cancel Subscription.

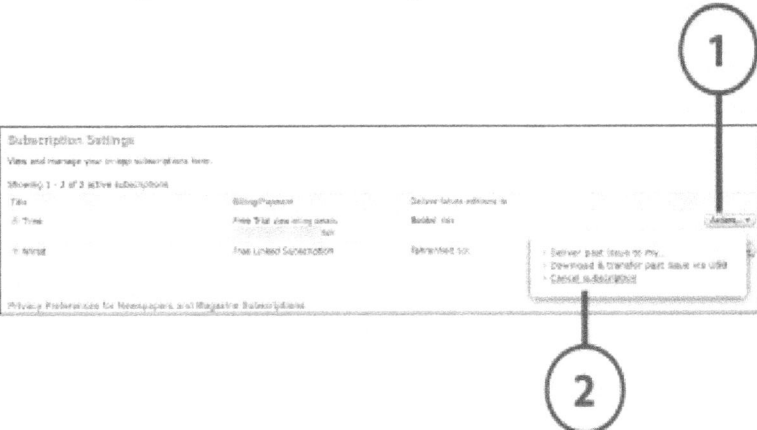

c. Choose one or more motives for canceling.

d. Input a comment if you select "Other".

e. Press cancel subscription.

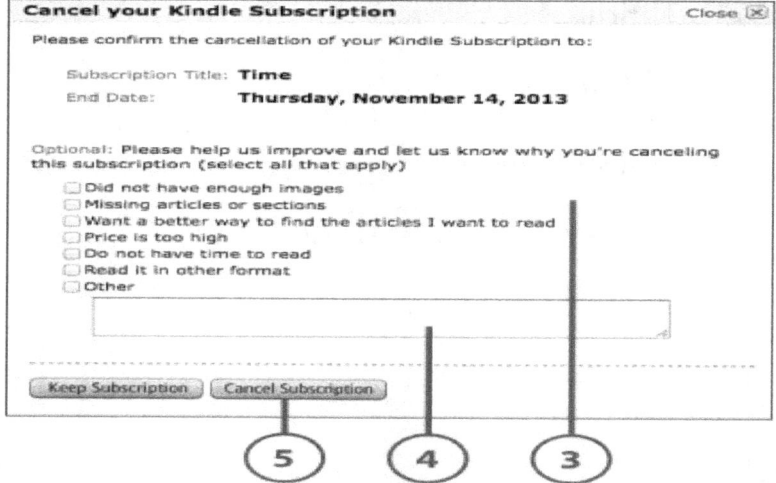

Way to Previous Problems for Canceled Subscriptions

If you decide to cancel a subscription, any problems that have been downloaded to your Kindle Fire will still be in your Kindle library. However, you won't be able to download any previous problems. Ensure downloading all the problems for which you paid for before you cancel.

REACTIVATING A CANCELED SUBSCRIPTION

Amazon keep up a list of all your inactive subscriptions. This list can be use to reactivate a canceled subscription without stress.

a. From subscription settings, tap view inactive subscriptions.

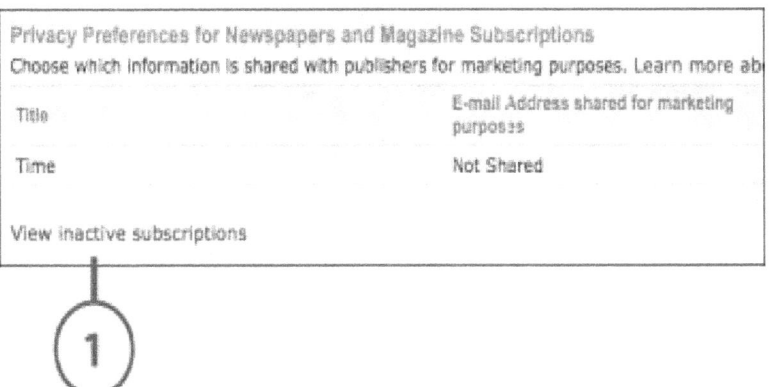

b. Tap "Actions" for the subscription you plan to reactivate.

c. Tap reactivate subscription.

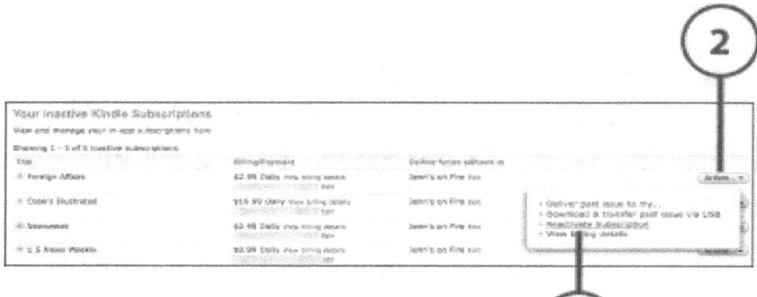

d. Again tap "Reactivate Subscription" in the confirmation dialog box.

Why Is the Action Button Missing?

If you deregistered your Kindle to which an active subscription was being delivered, then the actions button got lost. Before you will be able to reactivate the subscription, firstly you need to choose a Kindle to which the subscription should be delivered.

It's Not All Good: Resubscribing During a Free Trial

subscription commonly start with a free trial period, during which you get one or more problems. If nothing is done after the free trial, automatically you will be charged the month rate. However, during the free trial, if you cancel a subscription and then resubscribe, you are charged at once.

GETTING SUBSCRIPTION PRIVACY SETTINGS CHANGE

Be alert that Amazon doesn't share your email address with content providers unless you instruct them. That can be done using subscription privacy settings.

a. From subscription settings, search the privacy preference for Magazine and Newspaper Subscriptions.

b. Tap "Edit" for the subscription you plan to amend.

c. Verify the box(es) for the detail you plan to share with the content provider.

d. Again verify the suitable box, if you want your settings to be the default for future subscriptions.

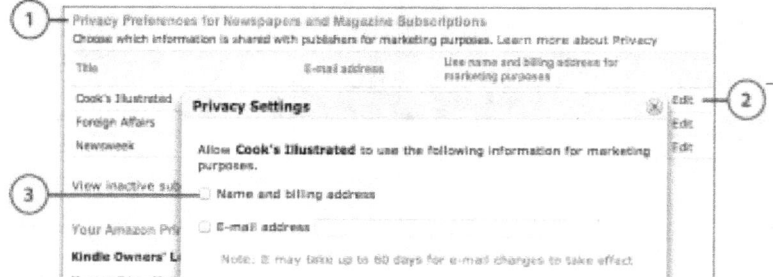

CHAPTER 4

KINDLE PAYMENT INFORMATION UPDATE

The credit card that Amazon uses for purchases and for present subscriptions can be change.

GETTING AMAZON PURCHASES CREDIT CARDS CHANGE

When you purchase Kindle books and MP3s, and rent or buy Amazon videos, the credit card been used for 1-click purchases at Amazon.com is billed automatically. However, the credit card detail can be change, put a new credit card, or by using Manage Your Kindle select a different credit card.

Several Credit Cards

amazon can save several cards for your account, and you can select which one is used for your 1-click purchases on your Manage Your Kindle page. Note that if you change your credit card it doesn't change the credit card that is been used for your subscriptions, ensure to change those separately.

   a. Tap "Kindle Payment Settings".

   b. Tap "Edit".

c. Input the details of your new credit card, or choose a different card.

d. Tap "Continue".

CHANGING PRESENT SUBSCRIPTION CREDIT CARD

Separately, you must ensure to update payment options for present subscriptions.

a. From Kindle Payment Settings, tap "Edit" for the subscription you plan to change

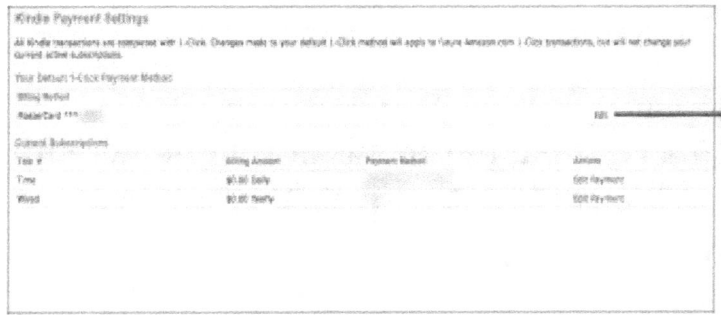

b. Input your new credit card detail

c. Tap "Continue".

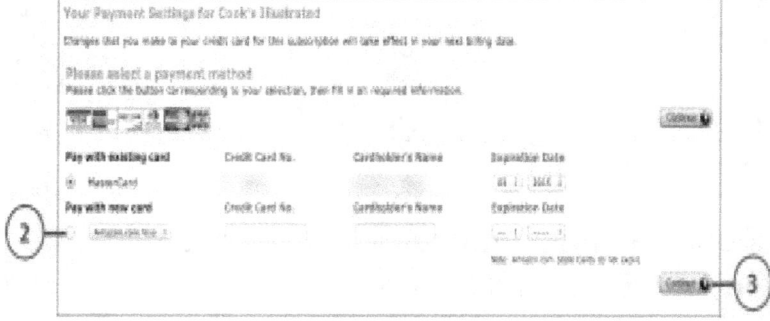

It's Not All Good: Updating Credit Card Detail

Whenever you change your payment detail, you have to recall to update each of your subscriptions. If you have several subscriptions, this can appear to be a time-consuming process, and it may be so easy to forget when you change your 1-click account.

CHAPTER 5

MANAGE YOUR KINDLE DEVICES

You can join several Kindles to your account. Having two or more Kindles registered to the same account is helpful when you and other members of your family have the same tastes in books. If you purchase a book on one Kindle, you will be able to read it on different Kindle at the same time without purchasing it again.

Therefore, the Manage Your Device page lists all your Kindle devices (including any Kindle apps installed on your phone, computer, or tablet). Your Kindle name can be change, and you can as well deregister a Kindle.

GETTING A KINDLE OR KINDLE APP DEREGISTER

If you make up your mind to give away or sell your Kindle Fire, you should first deregister it. This takes away the Kindle Fire's access to your account and stop the new owner from accessing or using your new credit card detail.

a. From the Manage Your Kindle page, tap "Manage Your Device".

b. Find the Kindle you plan to deregister and tap it.

c. Tap deregister.

d. On the pop-up of the deregister, tap the "Deregister Button".

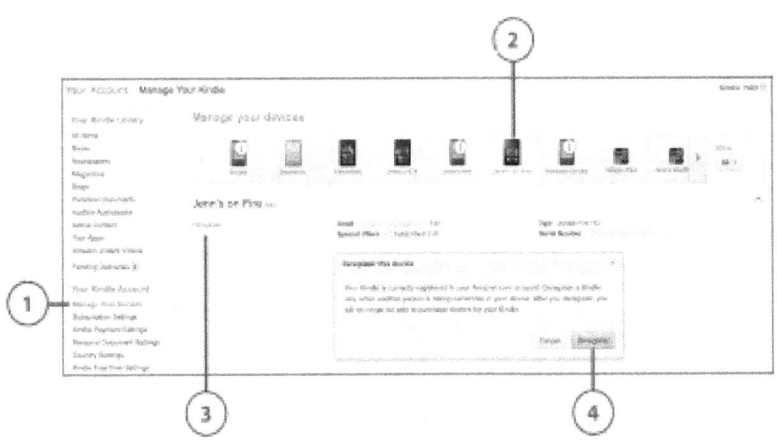

Why You Should Deregister an App?

If your phone, computer, or tablet got lost or stolen, or maybe you sold or give it away, you should ensure to deregister the Kindle app for that device. By doing so, no purchases can be made against your account without you been aware of it.

PLANNING TO RENAME YOUR KINDLE FIRE.

The name of your Kindle device can be change in order to make it appear special and be able to differentiate it from your other Kindles

a. From Manage Your Device, tap the Kindle whose name you plan to change.

b. Tap "Edit" next to the exiting name.

c. Input a new name for your Kindle.

d. Lastly, tap save.

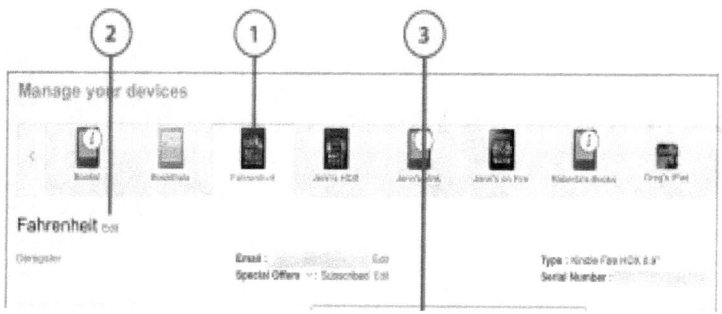

Naming Your Kindle Fire

Your device's name show in the left corner of the status bar. Unless you plan to have a look at something like "Jennifer's 2nd Kindle" every day, you may decide to change it.

SWITCHING OFF UNIQUE OFFERS

Your Kindle Fire shows unique offers and ads that show on the screensaver and in the offers heading of the navigation bar. You can pay a one-time fee to cancel these offers.

a. In Manage Your Device, find the that
   you plan to unsubscribe from the
   unique offers.

b. Tap "Edit" in the column of the
   unique offers.

c. Tap Unsubscribe Now with 1-click.
   Your account will be charged $15 to
   enable you unsubscribe other Kindle
   Devices varies.

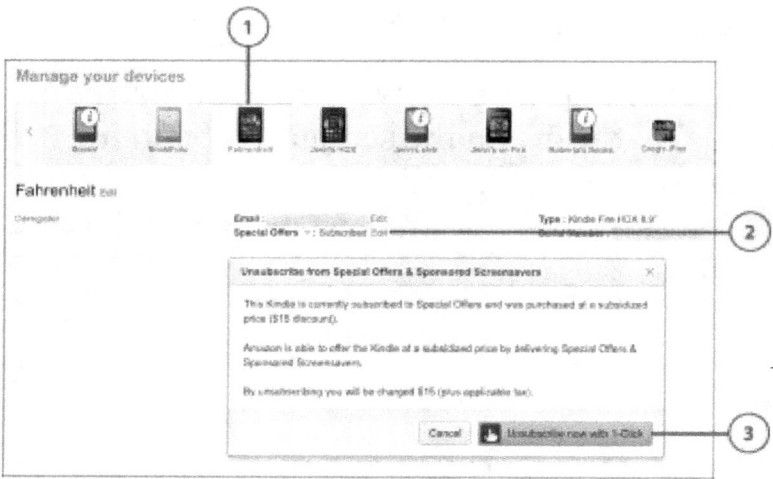

Why Should I Have to Pay?

Amazon gives charges for advertising fees for unique offers, which they claim subsidizes the cost of the Kindle Fire. Thus, if you opt out, their rationale is that they you to pay the projected difference in cost. Lot of people are not disturbed by the placement of the offers; in this case I advise you to make use of your Kindle Fire for a while to see whether removing them is worth the additional cost.

SWITCHING OFF WHISPERSYNC

Whispersync maintain all your devices and Kindle apps synchronized. It synchronizes the reading position, highlights, and notes of yours, and more. If you individually use several devices or apps when reading a book, ensure to switch on whispersync. If several people in your house read Kindles registered to the same account, you should disable whispersync to enable each device keeps a special page position, highlights, and notes for a book.

   a. From Manage Your Kindle, tap "Whispersync Device Synchronization".

   b. Tap "Switch Off or On" to toggle whispersync. At once this change takes effect.

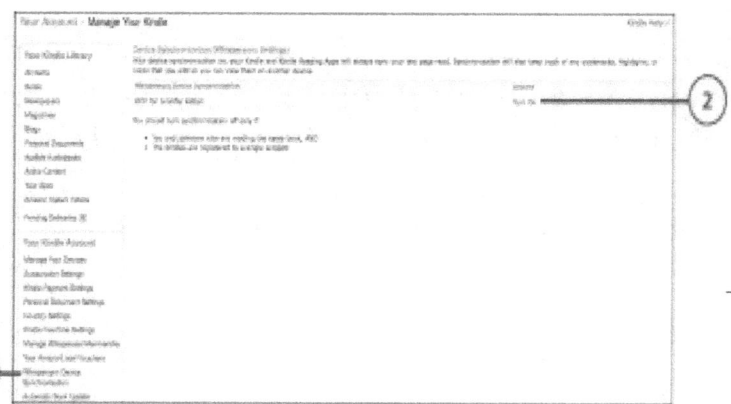

THE END